Dinosaurs
(And Other Prehistoric Animals)
Bingo Book

COMPLETE BINGO GAME IN A BOOK

Written By Rebecca Stark

ISBN 978-0-87386-442-8

Educational Books 'n' Bingo

Printed in the U.S.A.

DINOSAURS (& OTHER PREHISTORIC ANIMALS) BINGO DIRECTIONS

INCLUDED:

List of Terms

Templates for Additional Terms and Clues

2 Clues per Term

30 Unique Bingo Cards

Markers

1. **Either cut apart the book or make copies of ALL the sheets. You might want to make an extra copy of the clue sheets to use for introduction and review. Keep the sheets in an envelope for easy reuse.**

2. Cut apart the call cards with terms and clues.

3. Pass out one bingo card per student. There are enough for a class of 30.

4. Pass out markers. You may cut apart the markers included in this book or use any other small items of your choice.

5. Decide whether or not you will require the entire card to be filled. Requiring the entire card to be filled provides a better review. However, if you have a short time to fill, you may prefer to have them do the just the border or some other format. Tell the class before you begin what is required.

6. There are 50 terms. Read the list before you begin. If there are any terms that have not been covered in class, you may want to read to the students the term and clues before you begin.

7. There is a blank space in the middle of each card. You can instruct the students to use it as a free space or you can write in answers to cover terms not included. Of course, in this case you would create your own clues. (Templates provided.)

8. Shuffle the cards and place them in a pile. Two or three clues are provided for each term. If you plan to play the game with the same group more than once, you might want to choose a different clue for each game. If not, you may choose to use more than one clue.

9. Be sure to keep the cards you have used for the present game in a separate pile. When a student calls, "Bingo," he or she will have to verify that the correct answers are on his or her card AND that the markers were placed in response to the proper questions. Pull out the cards that are on the student's card keeping them in the order they were used in the game. Read each clue as it was given and ask the student to identify the correct answer from his or her card.

10. If the student has the correct answers on the card AND has shown that they were marked in response to the *correct questions,* then that student is the winner and the game is over. If the student does not have the correct answers on the card OR he or she marked the answers in response to *the wrong questions,* then the game continues until there is a proper winner.

11. If you want to play again, reshuffle the cards and begin again.

Have fun!

TERMS INCLUDED

adaptation

Allosaurus

Ankylosaurus

Apatosaurus

Archaeopteryx

biped

bird-hipped

Brachiosaurus

carnivore

Coelophysis

conifers

Cretaceous period

date

Diplodocus

eggs

Eohippus

era(s)

extinct

fossil(s)

geology

hadrosaurs

herbivore

herds

ichthyosaurs

Iguanodon

Jurassic

K-T boundary

lizard-hipped

mammal(s)

Mesozoic

omnivore(s)

Oviraptor

paleontologist

Paleozoic

Pangaea

periods

plesiosaurs

predator

prehistoric

prey

pterosaurs

quadruped

sauropods

Smilodon

Stegosaurus

taxonomy

Triceratops

trilobites

Tyrannosaurus rex

woolly mammoth

Additional Terms

Choose as many terms as you would like and write them in the squares.
Repeat each as desired. Cut out the squares and randomly
distribute them to the class.
Instruct the students to place the square on the center space of their card.

Clues for
Additional Terms

Write two or three clues for each new term.

_____ 1. 2. 3.	_____ 1. 2. 3.
_____ 1. 2. 3.	_____ 1. 2. 3.
_____ 1. 2. 3.	_____ 1. 2. 3.

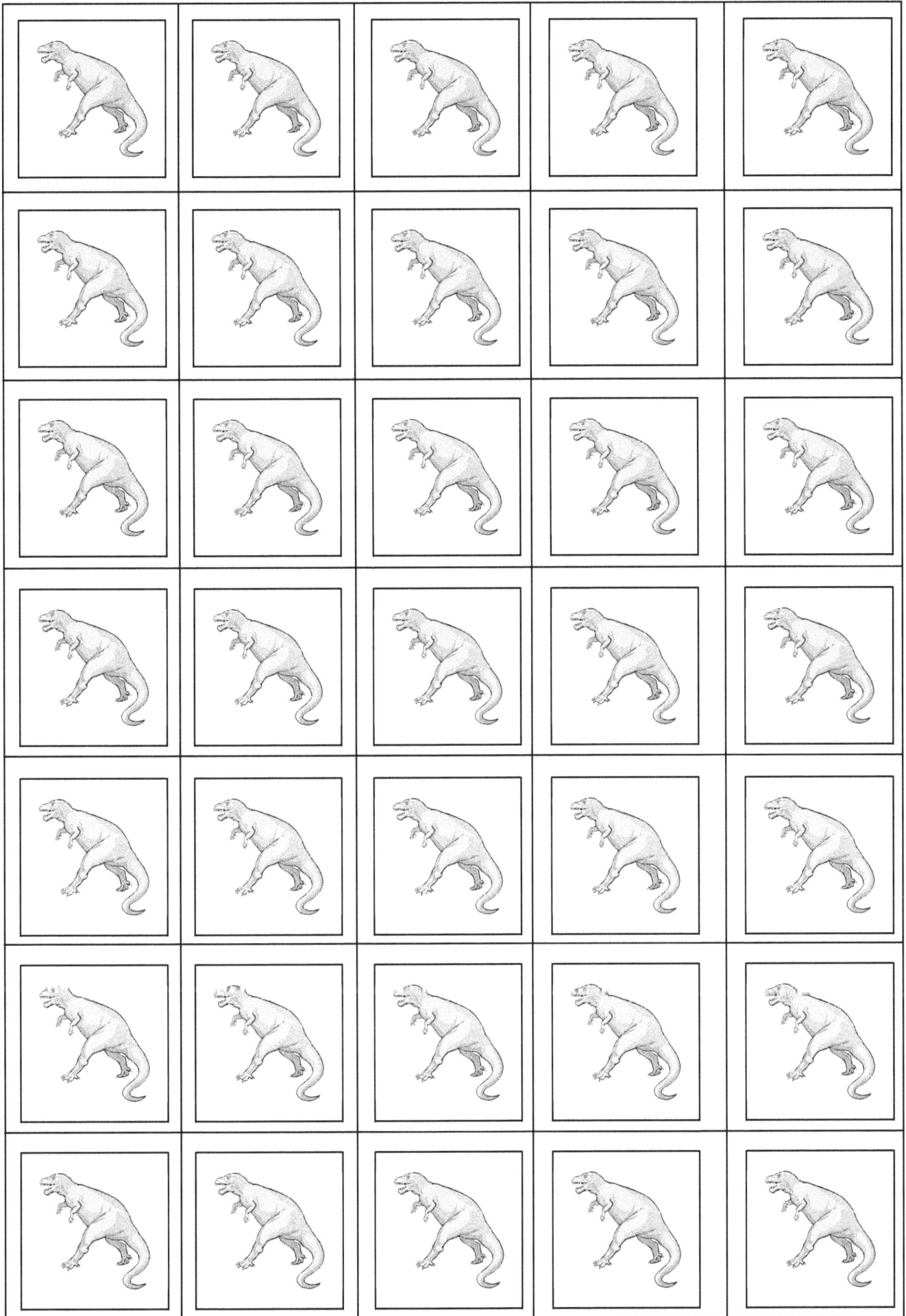

adaptation	Allosaurus
1. It is the response of an organism to changes in its environment. 2. ___ makes an organism become better suited to the habitat in which it lives.	1. This huge meat-eating dinosaur lived during the late Jurassic period. Like T-Rex and other theropods, it had small arms and a large head. 2. Like Tyrannosaurus Rex, this carnivorous theropod had strong hind legs.
Ankylosaurus	**Apatosaurus**
1. This armored plant-eater from the late Cretaceous period had a clubbed tail. 2. The name of this quadrupedal plant-eater comes from two Greek words: *ankulos,* meaning "curved or crooked," and *sauros,* meaning "lizard."	1. Its name means "deceptive lizard," but this gigantic plant-eater used to be called Brontosaurus, or "thunder lizard." 2. ___ was related to Diplodocus. ___ was not as long as Diplodocus, but it was bulkier. An adult may have weighed 33 tons!
Archaeopteryx	**biped**
1. It had feathers and could fly, but its structure was similar to that of the dinosaurs in many ways. 2. Although it was one of the earliest known birds, scientists do not think modern birds evolved from it.	1. A ___ is any animal that has two feet and walks upright. 2. Complete this analogy: quadruped : four :: ___ : two.
bird-hipped	**Brachiosaurus**
1. There were two main orders of dinosaurs: Saurischia, which comprised lizard-hipped dinosaurs, and Ornithischia, which comprised ___ dinosaurs. 2. Ornithischians were ___ dinosaurs, but birds did not evolve from this order.	1. Its name comes from two Greek words: *brachion,* meaning "arm," and *sauros,* meaning "lizard." 2. ___ was named this because its forelimbs were longer than its hind limbs. It had a relatively short tail.
carnivore	**Coelophysis**
1. Another word for meat-eater is ___. 2. A ___ eats herbivores and sometimes other ___s.	1. This small, carnivorous dinosaur lived during the late Triassic period. It had light, hollow bones, and its name means "hollow form." 2. This early predator was only about 9 feet tall. It walked quickly on its two legs.

conifers 1. Redwoods, yews, pines and other ___ were the dominant plant life during the Mesozoic Era. 2. ___ were the most important food source for the herbivorous dinosaurs.	**Cretaceous period** 1. The ___ marked the end of the Age of Dinosaurs. Tyrannosaurus Rex lived during this period. 2. At the end of the ___ many species of animals became extinct, including many of the dinosaurs.
date 1. Stratigraphy is often used to ___ a fossil. Paleontologists determine how deep the fossil is buried. 2. Carbon-14 can be used to ___ fossils up to about 60,000 years of age. It can be used to ___ ice-age mammals but not dinosaurs.	**Diplodocus** 1. ___ had a very long neck and a long, whip-like tail. Its hind legs were longer than its front legs. 2. ___ was a huge sauropod from the Jurassic period. Although not the heaviest land animal, it may have been the longest.
eggs 1. Baby dinosaurs hatched from ___. 2. Dinosaur ___ had hard, brittle shells.	**Eohippus** 1. ___ was the earliest known horse. It was only about 2 feet long and about 1 foot in height. 2. Its name means "dawn horse." It is also called Hyracotherium.
era(s) 1. A geological ___ is made up of two or more periods. Two or more ___ make up an eon, the largest division of geological time. 2. Dinosaurs first appeared in the Mesozoic ___. Humans first appeared in the Cenozoic ___.	**extinct** 1. We say that dinosaurs are ___ because they no longer exist. 2. A species that is in danger of becoming ___ is said to be endangered.
fossil(s) 1. The remains of a plant or animal found in stratified rock is called a ___. 2. A scientist who studies ___ is called a paleontologist.	**geology** 1. ___ is the study of the Earth and the materials of which it is made. 2. ___ involves the study of Earth's materials, structure and processes and how they have changed over time.

hadrosaurs 1. ___ are sometimes called "duck-billed" dinosaurs. 2. Some of these bipedal, herbivorous, duck-billed dinosaurs had crests on their heads.	**herbivore** 1. Another word for plant-eater is ___. 2. Complete this analogy: carnivore : meat :: ___ : plants
herds 1. Many herbivorous dinosaurs, such as Protoceratops, traveled in ___. 2. The reason many herbivorous dinosaurs traveled in ___ was probably for protection against predators.	**ichthyosaurs** 1. These aquatic reptiles looked a lot like modern-day porpoises. 2. The name of these prehistoric marine reptiles comes from two Greek words: *ichthyos,* meaning "fish," and *sauros,* meaning "lizard."
Iguanodon 1. The name of this bird-hipped herbivore means "iguana tooth." Its beak was toothless, but it had tightly packed cheek teeth. 2. This plant-eater had a horny beak. It also had a conical thumb spike on each thumb.	**Jurassic** 1. It is the middle period of the Mesozoic era. It started 206 million years ago and lasted until 144 million years ago. 2. Although dinosaurs first appeared in the Triassic period, the ___ period saw the rise of dinosaurs and the first birds.
K-T boundary 1. It stands for the transition from the Cretaceous period to the Tertiary period. 2. This transitional time was characterized by the extinction of many species of plants and animals, including dinosaurs.	**lizard-hipped** 1. There were two main orders of dinosaurs: Saurischia, which comprised ___ dinosaurs, and Ornithischia, which comprised bird-hipped dinosaurs. 2. Saurischians were ___ dinosaurs.
mammal(s) 1. Smilodon, a saber-toothed cat, and Eohippus, the earliest known horse, were prehistoric ___. 2. The first true ___ appeared in the early Jurassic period.	**Mesozoic** 1. The ___ Era is divided into the Triassic, the Jurassic, and the Cretaceous periods. 2. The name of this era means "middle animals."

Dinosaurs & Other Prehistoric Animals Bingo

omnivore(s)	Oviraptor
1. An ___ eats both plants and animals as its primary food source.	1. The name of this small, ominvorous dinosaur means "egg robber."
2. Most humans are called ___ because their primary food source comprises plants as well as other animals.	2. Paleontologists now think this dinosaur was wrongly named. They think the ___ fossil found with some fossilized eggs was probably the parent and not a predator that had stolen the eggs.

paleontologist	Paleozoic
1. This kind of scientist studies life of past geological periods.	1. The ___ era lasted from 543 million years ago to 248 million years ago. It began with the Cambrian period and ended with the Permian.
2. This kind of scientist studies the fossil remains of plant and animal life.	2. The ___ Era was followed by the Mesozoic Era.

Pangaea	periods
1. It is the name given to the supercontinent that existed before the continents separated.	1. Geological eras are divided into ___.
2. Scientists believe that ___ began to break up during the Triassic period.	2. The Mesozoic Era is divided into three ___: the Triassic, the Jurassic, and the Cretaceous.

plesiosaurs	predator
1. ___ were long-necked marine reptiles of the Mesozoic Era. They had four flippers and short, pointed tails.	1. An animal that hunts other animals for food is called a ___.
2. Like other ___, Elasmasaurus was a long-necked marine reptile with paddle-like flippers, a small head, and a pointed tail.	2. Allsosaurus was a top ___ in its food chain.

prehistoric	prey
1. It describes something that existed before recorded history.	1. An animal that is hunted by another animal for food is that animal's ___.
2. Dinosaurs, plesiosaurs, and pterosaurs are examples of ___ animals.	2. A predator feeds on its ___.

pterosaurs	**quadruped**
1. These flying reptiles first appeared in the late Triassic period and lived through most of the Cretaceous period. 2. Like other ___, pterodactyls had wings which were covered by a thin, leathery membrane.	1. A ___ is any animal that walks on four feet. 2. Complete this analogy: quadruped : ___ :: biped : two.
sauropods	**Smilodon**
1. The group of saurischians called the ___ included the largest land animals that ever lived. 2. The ___ were gigantic, quadrupedal, herbivorous dinosaurs. Diplodocus and Brachiosaurus are examples.	1. ___ was the largest of the saber-toothed cats. 2. Sometimes called the saber-toothed tiger, this prehistoric mammal was a fierce predator.
Stegosaurus	**taxonomy**
1. This plant-eater had armored plates along its back and a spiked tail. 2. The name of this bird-hipped dinosaur comes from two Greek words: *stegos,* meaning "cover or roof," and *saurus,* meaning "lizard."	1.The orderly classification of plants and animals is called ___. 2. Carolus Linnaeus is called the "father of ___. " He created the system for naming species that is used by biologists.
Triceratops	**trilobites**
1. This 3-horned plant-eater was among the last dinosaurs to roam the Earth before the great extinction 65 million years ago. 2. It had one horn above its beak and two longer horns above its eyes. A bony plate, called a frill, projected from its skull.	1. ___ are extinct marine arthropods of the Paleozoic Era. 2. The exoskeleton of these extinct marine arthropods was divided into three sections.
Tyrannosaurus rex 1. Like Allosaurus and other theropods, this carnivorous dinosaur had a large head and small forelimbs. 2. Like Allosaurus and other theropods, this carnivorous dinosaur of the late Cretaceous period walked on two legs.	**woolly mammoth** 1. This extinct mammal lived during the last Ice Age in the colder regions of the northern hemisphere. 2. This extinct animal resembled an elephant.

Dinosaurs & Other Prehistoric Animals Bingo

Dinosaurs & Other Prehistoric Animals Bingo

fossil(s)	Ankylosaurus	herds	Triceratops	quadruped(s)
carnivore	extinct	Stegosaurus	lizard-hipped	hadrosaurs
prehistoric	paleontologist		herbivore	Mesozoic
taxonomy	Allosaurus	era(s)	woolly mammoth	ichthyosaurs
Iguanodon	Tyrannosaurus rex	conifers	adaptation	geology

Dinosaurs & Other Prehistoric Animals Bingo

Triceratops	pterosaurs	Jurassic	Oviraptor	Iguanodon
ichthyosaurs	eggs	Brachiosaurus	Allosaurus	predator
Pangaea	Tyrannosaurus rex		Cretaceous period	era(s)
lizard-hipped	periods	paleontologist	trilobites	hadrosaurs
geology	Stegosaurus	conifers	carnivore	adaptation

Dinosaurs & Other Prehistoric Animals Bingo

Triceratops	era(s)	lizard-hipped	woolly mammoth	prehistoric
Tyrannosaurus rex	Ankylosaurus	biped	extinct	mammal(s)
Allosaurus	Stegosaurus		plesiosaurs	Apatosaurus
paleontologist	Pangaea	Iguanodon	eggs	Jurassic
adaptation	conifers	carnivore	trilobites	herds

Dinosaurs & Other Prehistoric Animals Bingo

paleontologist	plesiosaurs	herds	conifers	Iguanodon
K-T boundary	eggs	extinct	Oviraptor	prehistoric
herbivore	Brachiosaurus		quadruped	woolly mammoth
era(s)	Eohippus	Stegosaurus	carnivore	biped
adaptation	geology	omnivore(s)	Diplodocus	Mesozoic

Dinosaurs & Other Prehistoric Animals Bingo

geology	quadruped	Allosaurus	Brachiosaurus	conifers
K-T boundary	era(s)	biped	paleontologist	date
pterosaurs	Mesozoic		Ankylosaurus	herds
hadrosaurs	plesiosaurs	fossil(s)	trilobites	Diplodocus
lizard-hipped	carnivore	Paleozoic	Cretaceous period	herbivore

Dinosaurs & Other Prehistoric Animals Bingo

Apatosaurus	plesiosaurs	Jurassic	pterosaurs	Mesozoic
woolly mammoth	Allosaurus	Diplodocus	extinct	prehistoric
Oviraptor	biped		Brachiosaurus	Cretaceous period
carnivore	Iguanodon	trilobites	omnivore(s)	herbivore
ichthyosaurs	era(s)	fossil(s)	Paleozoic	herds

Dinosaurs & Other Prehistoric Animals Bingo

fossil(s)	plesiosaurs	prey	date	lizard-hipped
ichthyosaurs	herds	Tyrannosaurus rex	Ankylosaurus	prehistoric
Jurassic	woolly mammoth		Cretaceous period	Archaeopteryx
paleontologist	eggs	K-T boundary	Triceratops	Pangaea
conifers	carnivore	trilobites	omnivore(s)	Apatosaurus

Dinosaurs & Other Prehistoric Animals Bingo

herbivore	plesiosaurs	bird-hipped	woolly mammoth	Archaeopteryx
K-T boundary	pterosaurs	Oviraptor	herds	quadruped
prehistoric	predator		Mesozoic	Brachiosaurus
adaptation	paleontologist	Triceratops	Diplodocus	eggs
Stegosaurus	carnivore	omnivore(s)	Allosaurus	ichthyosaurs

Dinosaurs & Other Prehistoric Animals Bingo

Cretaceous period	lizard-hipped	Tyrannosaurus rex	prehistoric	Mesozoic
Diplodocus	pterosaurs	herbivore	Allosaurus	herds
mammal(s)	fossil(s)		Ankylosaurus	bird-hipped
Archaeopteryx	geology	Iguanodon	date	prey
eggs	trilobites	biped	Triceratops	quadruped

Dinosaurs & Other Prehistoric Animals Bingo

taxonomy	Triceratops	Brachiosaurus	Oviraptor	Paleozoic
Mesozoic	Archaeopteryx	extinct	Ankylosaurus	herds
plesiosaurs	predator		woolly mammoth	Pangaea
Iguanodon	hadrosaurs	Diplodocus	trilobites	mammal(s)
Coelophysis	geology	Jurassic	ichthyosaurs	herbivore

Dinosaurs & Other Prehistoric Animals Bingo

Apatosaurus	predator	Allosaurus	Diplodocus	ichthyosaurs
bird-hipped	mammal(s)	date	Cretaceous period	extinct
K-T boundary	pterosaurs		Jurassic	Tyrannosaurus rex
Coelophysis	prehistoric	trilobites	carnivore	Triceratops
biped	conifers	fossil(s)	omnivore(s)	lizard-hipped

Dinosaurs & Other Prehistoric Animals Bingo

lizard-hipped	eggs	mammal(s)	woolly mammoth	Cretaceous period
Tyrannosaurus rex	Stegosaurus	pterosaurs	omnivore(s)	K-T boundary
fossil(s)	prey		Mesozoic	Oviraptor
conifers	quadruped	herds	Triceratops	Ankylosaurus
predator	bird-hipped	plesiosaurs	biped	Archaeopteryx

Dinosaurs & Other Prehistoric Animals Bingo

Coelophysis	quadruped	Apatosaurus	mammal(s)	Mesozoic
pterosaurs	bird-hipped	plesiosaurs	Cretaceous period	Pangaea
woolly mammoth	Brachiosaurus		Tyrannosaurus rex	prey
herbivore	trilobites	Archaeopteryx	predator	Triceratops
carnivore	hadrosaurs	omnivore(s)	fossil(s)	date

Dinosaurs & Other Prehistoric Animals Bingo

conifers	pterosaurs	Allosaurus	Cretaceous period	Coelophysis
Archaeopteryx	fossil(s)	mammal(s)	Ankylosaurus	Pangaea
Diplodocus	woolly mammoth		Jurassic	Brachiosaurus
hadrosaurs	trilobites	plesiosaurs	biped	Apatosaurus
carnivore	Oviraptor	predator	ichthyosaurs	herbivore

Dinosaurs & Other Prehistoric Animals Bingo

date	Cretaceous period	Allosaurus	lizard-hipped	herds
Apatosaurus	Paleozoic	extinct	pterosaurs	Diplodocus
Mesozoic	fossil(s)		prehistoric	woolly mammoth
carnivore	mammal(s)	bird-hipped	trilobites	Coelophysis
ichthyosaurs	eggs	omnivore(s)	Jurassic	Tyrannosaurus rex

© Barbara M. Peller

Dinosaurs & Other Prehistoric Animals Bingo

Brachiosaurus	Smilodon	bird-hipped	Paleozoic	periods
Oviraptor	predator	prey	K-T boundary	taxonomy
Coelophysis	quadruped		Mesozoic	Tyrannosaurus rex
paleontologist	eggs	carnivore	date	Triceratops
Diplodocus	mammal(s)	omnivore(s)	Archaeopteryx	Pangaea

Dinosaurs & Other Prehistoric Animals Bingo

Coelophysis	sauropods	Eohippus	mammal(s)	carnivore
date	Diplodocus	trilobites	woolly mammoth	prey
Cretaceous period	taxonomy		Smilodon	bird-hipped
geology	ichthyosaurs	herbivore	Allosaurus	Pangaea
Iguanodon	biped	lizard-hipped	Triceratops	quadruped

Dinosaurs & Other Prehistoric Animals Bingo

herds	plesiosaurs	Archaeopteryx	Diplodocus	Oviraptor
geology	Coelophysis	Allosaurus	Mesozoic	biped
Cretaceous period	Pangaea		Eohippus	Paleozoic
prodator	extinct	trilobites	taxonomy	Jurassic
Smilodon	mammal(s)	Iguanodon	sauropods	Apatosaurus

Dinosaurs & Other Prehistoric Animals Bingo

Mesozoic	Apatosaurus	mammal(s)	bird-hipped	predator
date	conifers	Paleozoic	lizard-hipped	taxonomy
sauropods	woolly mammoth		Ankylosaurus	extinct
Jurassic	Smilodon	Iguanodon	eggs	Eohippus
prehistoric	periods	ichthyosaurs	herbivore	omnivore(s)

Dinosaurs & Other Prehistoric Animals Bingo

predator	sauropods	taxonomy	mammal(s)	Ankylosaurus
Brachiosaurus	Tyrannosaurus rex	K-T boundary	Iguanodon	Oviraptor
quadruped	prey		paleontologist	Eohippus
geology	herbivore	adaptation	eggs	Smilodon
era(s)	Stegosaurus	periods	Triceratops	extinct

Dinosaurs & Other Prehistoric Animals Bingo

Apatosaurus	date	K-T boundary	mammal(s)	hadrosaurs
quadruped	Eohippus	Archaeopteryx	bird-hipped	fossil(s)
Pangaea	ichthyosaurs		sauropods	Allosaurus
Iguanodon	lizard-hipped	Smilodon	geology	herbivore
paleontologist	periods	omnivore(s)	Coelophysis	eggs

Dinosaurs & Other Prehistoric Animals Bingo

prehistoric	Jurassic	Eohippus	pterosaurs	Coelophysis
Oviraptor	taxonomy	herds	bird-hipped	Ankylosaurus
Archaeopteryx	woolly mammoth		fossil(s)	prey
Smilodon	geology	eggs	extinct	conifers
periods	biped	sauropods	Pangaea	K-T boundary

© Barbara M. Peller

Dinosaurs & Other Prehistoric Animals Bingo

Brachiosaurus	sauropods	lizard-hipped	pterosaurs	omnivore(s)
Apatosaurus	predator	ichthyosaurs	date	extinct
Jurassic	Coelophysis		adaptation	fossil(s)
Pangaea	Stegosaurus	Smilodon	biped	eggs
hadrosaurs	herbivore	periods	Iguanodon	Eohippus

© **Barbara M. Peller**

Dinosaurs & Other Prehistoric Animals Bingo

Brachiosaurus	predator	conifers	sauropods	bird-hipped
Mesozoic	omnivore(s)	K-T boundary	Oviraptor	fossil(s)
prey	Paleozoic		Coelophysis	Pangaea
hadrosaurs	adaptation	Smilodon	biped	quadruped
era(s)	paleontologist	periods	taxonomy	Stegosaurus

Dinosaurs & Other Prehistoric Animals Bingo

paleontologist	K-T boundary	sauropods	Allosaurus	Eohippus
extinct	hadrosaurs	date	Brachiosaurus	Ankylosaurus
quadruped	bird-hipped		adaptation	Smilodon
Paleozoic	geology	Stegosaurus	periods	taxonomy
omnivore(s)	conifers	Archaeopteryx	Diplodocus	era(s)

Dinosaurs & Other Prehistoric Animals Bingo

Eohippus	sauropods	adaptation	Oviraptor	Paleozoic
Jurassic	woolly mammoth	bird-hipped	predator	Brachiosaurus
hadrosaurs	Iguanodon		taxonomy	paleontologist
Coelophysis	pterosaurs	geology	periods	Smilodon
prey	Diplodocus	Allosaurus	Stegosaurus	era(s)

Dinosaurs & Other Prehistoric Animals Bingo

adaptation	Archaeopteryx	sauropods	predator	Tyrannosaurus rex
hadrosaurs	Jurassic	date	Smilodon	Ankylosaurus
trilobites	Stegosaurus		periods	paleontologist
Paleozoic	Apatosaurus	era(s)	K-T boundary	extinct
Coelophysis	taxonomy	Eohippus	prehistoric	prey

Dinosaurs & Other Prehistoric Animals Bingo

Mesozoic	plesiosaurs	Triceratops	sauropods	Archaeopteryx
Tyrannosaurus rex	Eohippus	adaptation	Iguanodon	taxonomy
Stegosaurus	Pangaea		Paleozoic	Oviraptor
prey	prehistoric	ichthyosaurs	periods	Smilodon
pterosaurs	Cretaceous period	Coelophysis	era(s)	hadrosaurs

Dinosaurs & Other Prehistoric Animals Bingo

Eohippus	plesiosaurs	Paleozoic	date	Cretaceous period
hadrosaurs	Iguanodon	K-T boundary	prey	prehistoric
quadruped	sauropods		Ankylosaurus	adaptation
Tyrannosaurus rox	geology	herds	periods	Smilodon
Brachiosaurus	bird-hipped	era(s)	Apatosaurus	Stegosaurus

Dinosaurs & Other Prehistoric Animals Bingo

conifers	sauropods	Oviraptor	Cretaceous period	Smilodon
extinct	Paleozoic	Jurassic	taxonomy	Ankylosaurus
era(s)	Stegosaurus		prey	K-T boundary
hadrosaurs	Apatosaurus	plesiosaurs	periods	adaptation
geology	lizard-hipped	biped	Eohippus	herds